CRAYOLA INSECT COLORS

Christy Peterson

Lerner Publications ◆ Minneapolis

Copyright © 2022 by Lerner Publishing Group, Inc.

All rights reserved. International copyright secured. No part of this book may be reproduced, stored in a retrieval system, or transmitted in any form or by any means—electronic, mechanical, photocopying, recording, or otherwise—without the prior written permission of Lerner Publishing Group, Inc., except for the inclusion of brief quotations in an acknowledged review.

© 2022 Crayola, Easton, PA 18044-0431. Crayola Oval Logo, Crayola, Serpentine Design, Denim, and Screamin' Green are registered trademark of Crayola used under license.

Official Licensed Product
Lerner Publications Company
An imprint of Lerner Publishing Group, Inc.
241 First Avenue North
Minneapolis, MN 55401 USA

For reading levels and more information, look up this title at www.lernerbooks.com.

Main body text set in Mikado.
Typeface provided by HVD.

Editor: Andrea Nelson **Designer:** Laura Otto Rinne
Lerner team: Sue Marquis

Library of Congress Cataloging-in-Publication Data

Names: Peterson, Christy, author.
Title: Crayola insect colors / Christy Peterson.
Description: Minneapolis : Lerner Publications, [2022] | Series: Crayola creature colors | Includes bibliographical references. | Audience: Ages 4-9 | Audience: Grades K-1 | Summary: "Invite readers to explore the exciting world of insects with the help of Crayola. They will learn all about different insect characteristics, habitats, diets, and life cycles in this engaging introduction" —Provided by publisher.
Identifiers: LCCN 2020044803 (print) | LCCN 2020044804 (ebook) | ISBN 9781728424521 (library binding) | ISBN 9781728431130 (paperback) | ISBN 9781728430416 (ebook)
Subjects: LCSH: Insects–Juvenile literature. | Colors–Juvenile literature.
Classification: LCC QL467.2 .P469 2022 (print) | LCC QL467.2 (ebook) | DDC 595.7–dc23

LC record available at https://lccn.loc.gov/2020044803
LC ebook record available at https://lccn.loc.gov/2020044804

Manufactured in the United States of America
1-49322-49438-12/30/2020

TABLE OF CONTENTS

Meet the Insects 4

What Is an Insect? 6

Insect Life Cycles 10

Many Colors of Insects 18

Where Insects Live 24

Many Colors.......... 28
Glossary 30
Learn More............. 31
Index......................32

MEET THE INSECTS

This colorful moth is an insect.

More than one million kinds of insects live on Earth.

INSECT FACTS

Scientists are finding new kinds of insects every year. There are many more to discover!

GARDEN TIGER MOTH

WHAT IS AN INSECT?

Insects come in many shapes and sizes. But all insects have a head, a thorax, and an abdomen. They have six legs.

INSECT FACTS

The longest insect in the world is 25 inches (64 cm) long. That's about the length of seven crayons!

Can you count this **black** ant's three body parts?

Insects have a hard outside covering called an exoskeleton. This beetle's exoskeleton is **green**, **indigo**, and **orange**!

JEWEL BEETLE

INSECT LIFE CYCLES

Most insects hatch from eggs. This **red** ladybug lays **yellow** eggs.

11

PIPEVINE SWALLOWTAIL CATERPILLAR

This **black**-and-orange caterpillar doesn't look like its parents.

Soon it will make a chrysalis. Inside it will change into a butterfly.

Some insect babies look like small adults. This **yellow-green** katydid will get wings when it grows up.

15

COLOR IN ACTION

This butterfly's striking green-and-yellow coloring signals to other butterflies that it is a male.

Can you find these colors in the photo?

QUEEN ALEXANDRA'S BIRDWING BUTTERFLY

MANY COLORS OF INSECTS

Some insects use color to help them hide from predators. This **brown** bug looks like a leaf!

LEAF MIMIC KATYDID

HONEYBEE

This bee may sting to protect itself. Its **black** and yellow stripes are a warning.

INSECT FACTS

This fly can't sting at all. But its stripes still keep predators away!

BUMBLEBEE HOVERFLY

COLOR IN ACTION

This butterfly's orange-and-black wings signal caution. "Don't eat me. I taste bad."

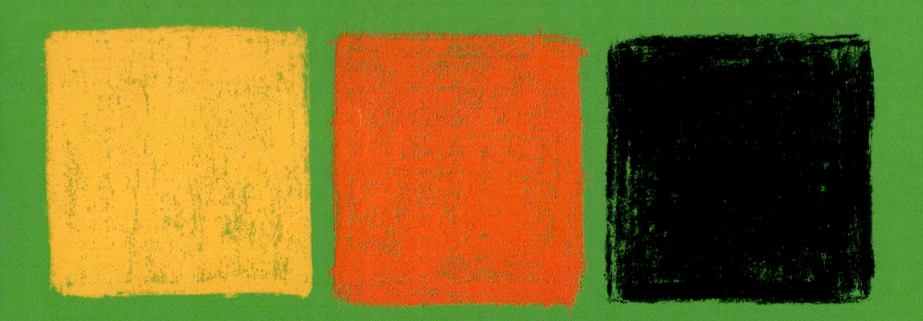

Can you find these colors in the photo?

22

MONARCH BUTTERFLY

WHERE INSECTS LIVE

Insects live all over the world. They live in hot places and cold places.

25

They live in dry places and wet places.

Look around! What insects do you see where you live?

GRASSHOPPER

LARGE MILKWEED BUG

ANT

PIPEVINE SWALLOWTAIL

LEAF INSECT

27

MANY COLORS

Insects come in lots of different colors. Here are some of the Crayola crayon colors used in this book.

RED

GOLDENROD

BLUE VIOLET

SCREAMIN' GREEN

DENIM

29

GLOSSARY

abdomen: the third section of an insect's body

caterpillar: a butterfly or moth in its larval stage

chrysalis: the protective casing around a butterfly as it changes from a larva to an adult

exoskeleton: the hard outside covering of an animal like a beetle, spider, or crab

hatch: to emerge from an egg

insect: an animal with six legs and three body parts

katydid: an insect related to grasshoppers and crickets

predator: an animal that eats other animals

thorax: the second section of an insect's body

warning: a signal or alarm that alerts the onlooker to danger

LEARN MORE

Gibson, Roberta. *How to Build an Insect*. Minneapolis: Millbrook Press, 2021.

Jenkins, Steve. *Insects: By the Numbers*. Boston: HMH Books for Young Readers, 2020.

Schuh, Mari. *Our Colorful World: A Crayola Celebration of Color*. Minneapolis: Lerner Publications, 2020.

INDEX

abdomen, 6

chrysalis, 13

egg, 10

exoskeleton, 8

thorax, 6

PHOTO ACKNOWLEDGMENTS

Image credits: Katja Schulz/flickr (CC BY 2.0), pp. 1, 15; Nick Goodrum/flickr (CC BY 2.0), p. 5; Macro Monster/Wikimedia Commons (CC BY-SA 2.0), p. 7; yod67/Shutterstock.com, pp. 9, 29 (top left); Jorge Abel Photography/Shutterstock.com, pp. 11, 28 (left); Sundry Photography/Shutterstock.com, p. 12; Ginger Wang/Shutterstock.com, p. 13; Russell Marshall/Shutterstock.com, p. 17; Dr Morley Read/Shutterstock.com, p. 19; Dancestrokes/Shutterstock.com, pp. 20, 28 (right); MR.AUKID PHUMSIRICHAT/Shutterstock.com, p. 21; Darkdiamond67/Shutterstock.com, p. 23; Tomasz Sowa/Shutterstock.com, p. 25; Benny Marty/Shutterstock.com, p. 26; akslocum/Shutterstock.com, p. 27 (top left); John Flannery/flickr (CC BY-ND 2.0), pp. 27 (top right), 29 (bottom); Pavel Kirillov/flickr (CC BY-SA 2.0), pp. 27 (bottom), 29 (top right).

Cover: Digital Images Studio/Shutterstock.com (front); Jorge Abel Photography/Shutterstock.com (back).